Home Office

Dealing with Disaster

LONDON: HMSO

ISBN 0 11 341044 1

Introductory Note

There are variations from one part of the United Kingdom to another in the roles and responsibilities of central and local government, in the offices held and in the terminology used within governmental and other organisations referred to in this guidance. To try to reflect all these variations would make the text much lengthier and considerably less readable. Accordingly, while much of the principle is universally applicable, the guidance reflects the arrangements and terminology commonly used in England and Wales.

Contents

Foreword by the Home Secretary

Disasters usually strike suddenly, unexpectedly and with ferocity. The response invariably involves the emergency services as part of their duties and it generally involves local authorities as an extension of their responsibilities.

A range of other organisations may also be involved, such as electricity, gas, water and telecom companies, manufacturing and transport industries. Experience indicates that we should not underestimate the pressure on those who respond. Not only should they know their own roles in the event of disaster, but they should also appreciate the way in which their roles interact with others. It is, therefore, prudent to prepare.

I commend this guidance, put together by my Civil Emergencies Adviser, as a contribution towards preparedness. I know that he has received considerable advice and assistance from many people in the emergency services, in local authorities and elsewhere, and would like to thank them all for their cooperation.

The guidance is not in any sense meant to be prescriptive nor does it replace existing single service guidance. The key message which it brings across is the need for the principal players in the disaster response to cooperate in its preparation – planning, training and exercising – and in the event. That is a sentiment which I strongly endorse and, if this guidance succeeds in getting that message more widely understood, then it will have served its purpose.

The Rt. Hon. Kenneth Clarke QC MP

1 Introduction

1.1 The purpose of this document is to assist those involved in the preparation for and response to disasters. It offers advice on good practice based on the experience of those who have had to handle disasters in the past. The guidance is designed to show any who might be involved how they can fit into the overall disaster response. It is intended to supplement and not to replace existing 'single-service' guidance on disaster preparedness and response which is available within many organisations.

1.2 The advice is general in nature and covers those topics which are of interest to all who might be called upon to respond. This includes organisations for whom emergencies are part and parcel of everyday life – police, fire, HM Coastguard, the NHS ambulance and emergency medical services and local authorities – through to a great many others for whom involvement is a more occasional matter but who may nevertheless be called upon to respond quickly when the need arises. These include central government departments, the armed forces, voluntary organisations and commercial and industrial concerns, including the utilities (water, gas, telephone and electricity companies).

1.3 Above all, this guidance is not intended to be prescriptive. It is meant to inform but not dictate the decisions that have to be taken either in the often chaotic situation following a disaster or in the more measured atmosphere of response preparation. No single organisational arrangement will be appropriate to each and every disaster, nor will a single organisational blueprint for planning meet every need. The key to an effective response is to apply sound basic principles to the problem at hand. Where examples of organisations or structures are offered in this guidance they must be seen as no more than examples. They are not, and should not be regarded as, the 'right' solution for all circumstances.

1.4 Disasters can have a variety of causes, such as:

Natural – Storms, floods, snowfall, landslides, forest fires, earthquakes.

Technological – Damage to structures and buildings; industrial (eg explosion, release of toxic or radioactive substances); transport accident (air, sea, inland water, rail, road).

Social – Health emergencies (epidemics), poisoning of the food chain.

Environmental – Air, water and land pollution or contamination.

Hostile Acts – Terrorist acts or actions of a hostile state.

1.5 A disaster is commonly understood by the general public as a great misfortune or calamity. In the context of emergency planning, a useful working definition of a disaster is any event (happening with or without warning) causing or threatening death or injury, damage to property or the environment or disruption to the community, which because of the scale of its effects cannot be dealt with by the emergency services and local authorities as part of their day-to-day activities.

1.6 This working definition encompasses the definitions used by the emergency services for a 'major incident'. Some of these are set out in Annex A to this guidance. In defining a major incident, the emergency services recognise that there will be a need for special arrangements to be brought into play to respond to such an incident. That the emergency services should use the term 'major incident' is right because their contingency arrangements and operational response must be geared to any situation beyond their normal day-to-day activities. However, not every major incident will become a disaster: for example, a serious accident on a motorway involving a large number of vehicles will be a major incident if it demands special arrangements on the part of the police, fire and the NHS (both ambulance and hospital-based services) but unless it has some wider ranging effects on the community it is unlikely to be regarded as a disaster.

1.7 While the causes of disaster may be sudden and unpredictable, certain kinds of industrial activity carry known risks and are subject to legal requirements for emergency planning. These include known chemical or nuclear hazards at fixed locations, where the most probable types of incident and their likely consequences are largely foreseeable. For this reason it is possible to make detailed plans in advance for the appropriate action to be taken. The existence of such plans reduces the likelihood of errors resulting from decisions being taken under crisis conditions. The requirements are contained in the Control of Industrial Major Accident Hazards (CIMAH) Regulations 1984 and in conditions attached to licences issued under the Nuclear Installations Act 1965 (which is expected to be supplemented by off-site nuclear emergency planning regulations). The response to emergencies where these requirements are relevant is not addressed in detail in this guidance.

1.8 Additionally, the guidance is directed mainly at disasters occurring on land and within the UK. Marine accidents and aviation accidents which occur over the sea or abroad are not addressed so far as the immediate response is concerned. But such disasters may have major ramifications within the UK (for example, in respect of investigation and of the post-disaster treatment of survivors, relatives and friends) where a substantial number of British subjects are involved. Indeed this may apply to any other disaster abroad. The guidance in the following chapters should therefore be kept in mind by those likely to have to deal with such cases.

1.9 Disasters have a variety of effects on society and the environment. Thus they demand a combined and co-ordinated response, linking the expertise and resources of the emergency services and the local authorities, supplemented as appropriate by other services and organisations. There is no single agency within the UK which has all the skills and resources which may be needed. Moreover, a Home Office review[1] concluded that the disaster response would not be helped by the creation of anything in the nature of a national disaster squad: prime responsibility for handling disasters should remain at the local level where the resources and expertise are found. This guidance is aimed primarily at assisting that local response.

---

[1]Home Office Review of Arrangements for Handling Major Civil Disasters in the UK. Hansard 15 June 1989 column 515–516.

2 The combined response

2.1 There can be no one ideal response to disaster; the response will vary just as the nature and effects of the disaster will vary. Nevertheless, any response must be a combined and co-ordinated operation and certain other features will be common to the response to many disasters. Some key features are addressed in this chapter:

a. The core of the initial response will normally be provided by the emergency services supported by the appropriate local authority or authorities and voluntary organisations. Depending on the circumstances, crucial parts may also be played by a range of agencies and organisations including any industrial or commercial organisation involved, other industrial or commercial concerns, utilities, armed forces and central government (see paragraphs 2.3–2.5).

b. The basic objectives of the combined and co-ordinated response will be similar on each occasion (see paragraph 2.6).

c. The same basic management structure will often be applicable (see paragraphs 2.7–2.31) and

d. There will be a need to ensure that essential records are kept for debriefings, formal inquiries and publicising lessons learnt (see paragraphs 2.32 and 2.33).

2.2 To be effective, the combined response must be prepared in advance through planning, training and exercising. Guidance on these activities is contained in Chapter 7.

Agencies providing or contributing to the local response

2.3 The initial response to a disaster is usually provided by the emergency services supported by the local authority but many agencies can become involved. The emergency services maintain a state of readiness so that they can provide a rapid response and alert local authorities and other services as soon as possible. All organisations who need to respond quickly to a disaster should have arrangements which can be activated at short notice. These arrangements should be clearly established and promulgated.

2.4 Single service guidance documents have already been produced by or for a number of services. Some of these documents are listed in Annex B. This chapter draws on them to offer guidance on how the procedures and operations of each of the organisations involved can be combined and co-ordinated to provide an efficient and effective response to disaster.

2.5 Each service or agency working at the scene of a disaster has its own role and functions:

a. *The police* co-ordinate the activities of all those responding at and around the scene, which must – unless a disaster has been caused by severe weather or other natural phenomena – be treated as the scene of a crime and preserved accordingly. They oversee any criminal investigation. They also facilitate inquiries carried out by the responsible accident investigation body, such as the Health and Safety Executive, Railway Inspectorate or the Air or Marine Accident Investigation Branch. The police process casualty information and have responsibility for identifying and arranging for the removal of the dead. In this task they act on behalf of *HM Coroner* who has the legal responsibility for investigating the cause and circumstances of deaths arising from a disaster.

b. The first concerns of the *fire service* are to rescue people trapped in wreckage or debris and to prevent further escalation of the disaster by tackling fires and/or dealing with released chemicals and other hazards.

c. The *ambulance service*, in conjunction with the medical incident officer and medical teams, seeks to save life through effective emergency treatment at the scene and by transporting the injured in order of priority to receiving hospitals.

d. The primary responsibility of *HM Coastguard* is to initiate and co-ordinate civil maritime search and rescue within the United Kingdom Search and Rescue Region. This includes mobilising, organising and despatching resources to assist people in distress at sea or in danger on the cliffs or shoreline.

e. In the immediate aftermath of a disaster, the principal concerns of *local authorities* include support for the emergency services, support and care for the local and wider community and co-ordination of the response by organisations other than the emergency services. As time goes on, and the emphasis switches to recovery, the local authority will take a leading role in rehabilitating the community and restoring the environment.

f. *Volunteers* can contribute to a wide range of activities, either as members of a voluntary organisation or as individuals. The valuable part which volunteers can play is addressed in Chapter 5.

g. *Industrial or commercial organisations, including the utilities*, may play a direct part in the response to disaster if their personnel, operations or services have been involved. Other industries or commercial organisations may provide support, for example by providing equipment, services or specialist knowledge.

h. *Military assistance* may be used in support of the civil authorities. This has been an important part of many disaster responses in the past. Resources and skills which may be available and methods of requesting military assistance are set out in 'Military Aid to the Civil Community: A Pamphlet for the Guidance of Civil Authorities and Organisations'. (Ministry of Defence 1989, third edition: AC 60421.)

i. *Central government* may provide advice or support. The role of central government is discussed in Chapter 6.

Objectives for a combined response

2.6   The roles and responsibilities just described must be seen in the context of the objectives of the disaster response. All services and agencies responding to a disaster will be working – notwithstanding their particular responsibilities – to these common objectives. They are:

a. to save life;

b. to prevent escalation of the disaster;

c. to relieve suffering;

d. to safeguard the environment;

e. to protect property;

f. to facilitate criminal investigation and judicial, public, technical or other inquiries; and

g. to restore normality as soon as possible.

The response to a disaster at a single site

2.7   Within the UK, there is ample experience of disasters occurring within the bounds of relatively small areas. Many of the principles which emerge can also be applied to more widespread disasters, as discussed in paragraphs 2.25–2.31.

2.8   The scene immediately after disaster has struck is likely to be chaotic. Survivors may be helping each other and those who happen to be nearby may also be assisting. To bring some order to the chaos, it is important that the emergency services establish control over the immediate area and also build up arrangements for co-ordinating the contributions to the response. Experience has shown that an effective response depends on sound decisions being made and appropriate actions set in train at the outset.

2.9   It is generally accepted that the first member of the emergency services to arrive on the scene should not immediately become involved with rescue but should make a rapid assessment of the disaster and report to that service's control. Such information as is immediately available should be provided about the nature of the disaster and its location; the number of dead, injured and uninjured; hazards actual and potential; access to the site and possible rendezvous points; and which emergency services are present or required. Additionally, each of the emergency services has its own requirements; for example, in the case of the fire service, the number of pumps and appliances likely to be needed.

2.10   The control which receives the initial message should immediately, and in accordance with established plans, alert the other emergency service control rooms, the local authority and (where appropriate) the commercial, industrial or other organisation(s) involved. Where a commercial organisation has raised the alarm and taken initial action, for example in the case of an air accident at or near an airport, the same procedures should be followed.

2.11   At the scene, it is vital that the emergency services establish control and co-ordination arrangements at the earliest stage. Each service needs to establish its own control arrangements but continuing liaison between the various controls throughout the response phase is essential. The underlying principle is that the police assume the role of overall co-ordination, thus enabling the other services to concentrate on their specific tasks.

2.12 A number of arrangements may have to be made, for example:

a. assigning the control of specific functions to one of the emergency services or other agency, taking account of the circumstances of the disaster, the professional expertise of the emergency services and other agencies and any statutory obligations;

b. setting up an inner cordon to control access to the immediate scene both to prevent interference with fire-fighting and rescue operations and to preserve evidence at the disaster site;

c. the location of a collection point for survivors before they are taken to a survivor reception centre; the location of a casualty clearing station to which the injured can be taken; and an ambulance loading point for those who need to be taken to hospital;

d. the location of a rendezvous point or points for the emergency service, volunteers and other non-emergency services personnel;

e. the positioning of emergency service and other vehicles when they arrive at the scene, traffic flow into and out of the site, the location of a marshalling area; and

f. the location of a media liaison point (see paragraph 4.6b).

2.13 The possible need for evacuation may also have to be addressed at a very early stage. Chapter 3 deals with this in more detail.

2.14 Where the response is likely to prove prolonged or complex, each service may establish management arrangements at the site to oversee the control and co-ordination of its response, often from specially designed vehicles equipped with suitable communications. Figure 1 shows an example of a typical organisation for dealing with disaster.

2.15 Again, liaison in the establishment of incident controls is vital. The location of each emergency service's incident control must depend on circumstances but they would normally be in the vicinity of the disaster site and experience of previous disasters has shown that it makes for better co-ordination if the incident controls are located together.

2.16 Two other issues may need to be addressed as soon as the situation permits. First, if practical, an outer cordon may have to be established around the scene of the disaster to control access to the whole of the disaster site. Secondly, emergency flying restrictions may be required. Downdraught and noise from helicopters can disrupt rescue and fire-fighting operations, cause hazards to fire-fighters and rescuers, damage property and destroy evidence. The presence of unauthorised aircraft poses collision hazards to police and rescue helicopters operating in the area. It may therefore be necessary to regulate flights in the vicinity of the disaster and the National Air Traffic Services (NATS) Instructions for Establishing Emergency Flying Restrictions within the United Kingdom 1989 describe how these restrictions are imposed.

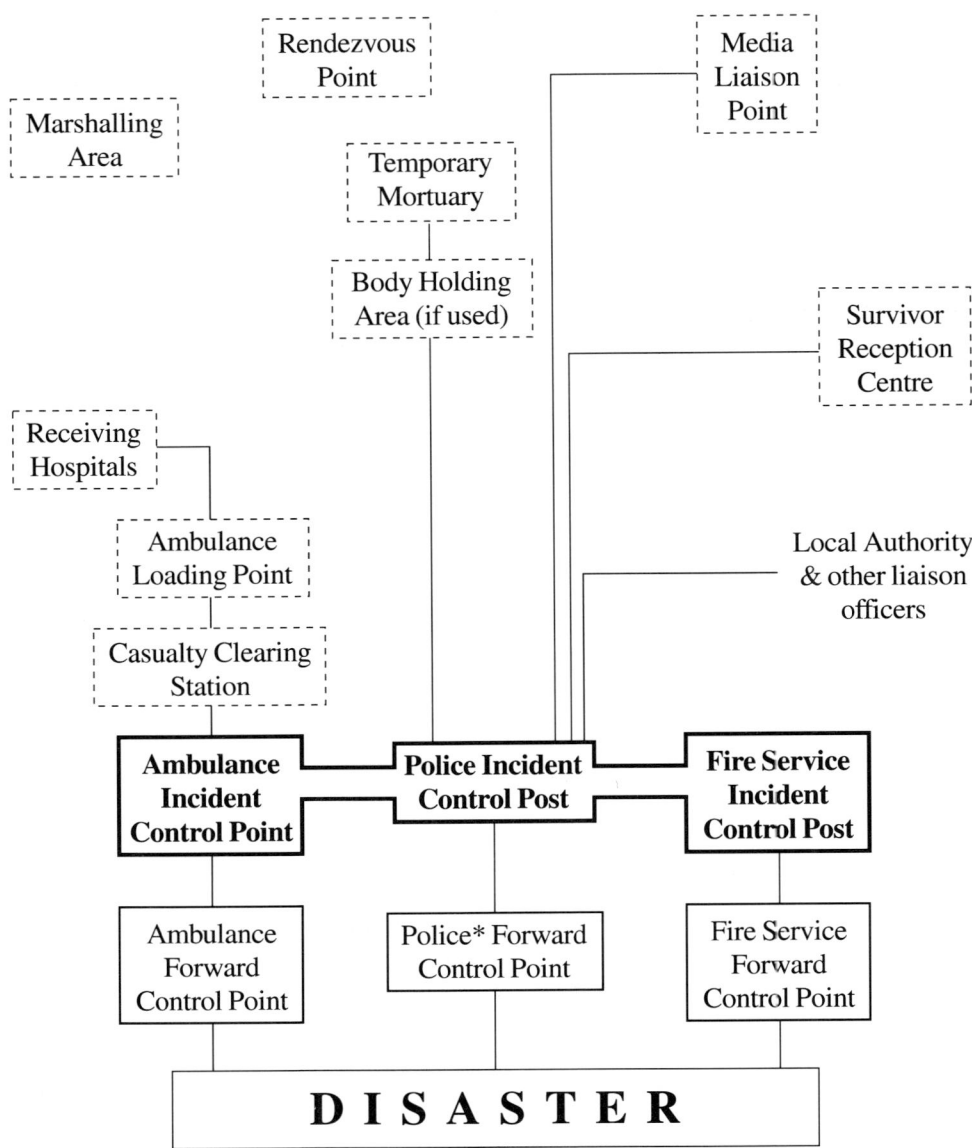

Figure 1 Example of a typical organisation for dealing with disaster

*May be withdrawn once Police Incident Control Post is established

2.17 It is good practice to plan for an officer who can represent the interests of the relevant local authorities to be present at the scene. This person should maintain contact with the incident controls as necessary. Arrangements should be in place for calling to the scene liaison officers from other organisations which may need to contribute to the disaster response, for example, HM Coastguard, the National Rivers Authority, gas, electricity or water companies or industrial or commercial concerns.

2.18 Liaison officers at the scene should be equipped with their own communications so that they can remain in contact with their own organisation to ensure that further support can be rapidly provided as necessary. Where local authority services might be required at short notice, resources should be mustered off-site so that they stand ready for immediate action if called upon by the emergency services.

2.19 If the disaster occurs within the perimeter of an industrial or other establishment it is advisable to appoint a site incident officer from the affected organisation to ease access to facilities within the establishment and act as a link to the establishment's senior management and emergency management structure.

Maritime disasters

2.20 If there are no standing arrangements already in place, the response to a maritime disaster will be co-ordinated by the appropriate Maritime Rescue Co-ordination Centre. However, reception of survivors ashore is the responsibility of the land-based authorities. It is essential that land and maritime authorities liaise at the earliest opportunity to tackle the particular problems caused by maritime disasters, including wreckage and pollution.

Senior management arrangements

2.21 Disasters can place considerable demands on the resources of the organisations called upon to respond, with consequent disruption to day to day activities. They may have long-term implications for people or the environment. Such matters require attention by senior management and, in local authorities, by elected members. The issues which arise will undoubtedly affect the responsibilities or activities of more than one organisation and experience has shown that such issues can best be addressed by establishing a co-ordinated approach at senior management level as well as at the scene.

2.22 The precise arrangements for putting the co-ordinated approach into effect at a senior level must depend on the nature of the disaster. But with a single site disaster, such as a train crash, one approach is to establish a single senior co-ordinating group based on the planning group described in paragraph 7.2. Whatever the arrangements, they need to be put in place quickly.

2.23 The emergency services and the local authorities are likely to be involved in the arrangements in the great majority of cases. It is also good practice to ensure that the needs of the media are addressed at this senior level, and where a co-ordinating group is established it should include a public relations manager whose role is described in paragraph 4.14. Other organisations which have a significant contribution to make to the disaster response should also be included, for example specialist police forces, commercial organisations, utilities, central government and the military authorities if military assistance is required. It is important that members of the group have the authority to make decisions and commit resources.

2.24 The purpose of the corporate approach at senior level is essentially to take strategic decisions in relation to the disaster response. Typical issues which may need to be addressed include the adequacy of resources for dealing with the disaster and support for those working at the scene; care of those dealing with or affected by the disaster including the relatives and friends of victims; longer-term implications for the community or the environment; relations with the media; visits by VIPs; and contacts with central government.

Handling widespread disasters

2.25 Historically the UK has been more fortunate than some other countries, suffering less from widespread disasters caused, for example, by storms, flooding or earthquakes. There is enough experience, however, to highlight some characteristics of the response to widespread disaster.

2.26 In a generally densely populated country, where widespread disasters are likely to affect large numbers of people, self-help is likely to be the first response. The disaster will probably overwhelm local resources, disrupt telecommunications and block the usual access routes. Any organised response to a widespread disaster may well start at a local–shire district or metropolitan borough–level, or at parish level in rural areas. In all cases the response would probably be supported by the relevant voluntary organisations.

2.27   It may be some hours before the scale and extent of a widespread disaster can be established and recovery plans activated. The management arrangement will need to be adapted to the task in hand and should be flexible enough to adapt to changing circumstances, but good practice is again to adopt a corporate approach using the principles already described in paragraphs 2.21–2.24. The precise arrangements, including which organisation should take the lead, will be determined by the nature of the disaster and its geographic extent. It may, for example, be necessary to establish co-ordination arrangements at several levels, national, regional, county and below.

2.28   Where the most severe, localised damage has occurred, for example, if residential buildings in different areas have collapsed, a single site might be managed along the lines suggested in paragraphs 2.7–2.19 and a number of sites might be managed by a single co-ordinating group along the lines set out in paragraphs 2.21–2.24.

2.29   Arrangements for responding to the complexities of a widespread disaster are difficult to devise under pressure in the early stages. Planning, involving the identification of possible disasters and the appropriate response, needs to address the kind of management structures which would be appropriate.

2.30   Resources will almost certainly be stretched by a widespread disaster. The authorities may have to call on the resources which industry and the armed forces can provide. Such resources can be mobilised more quickly if their availability and means of obtaining them have also been identified in advance. The role of central government in these and other circumstances is addressed in Chapter 6.

2.31   As already noted, there is no one model for dealing with the complex problems which widespread disasters pose and there has been little recent experience of them in the UK on which to base guidance. But the basic principles do not change and a short account of the initial response to the East Coast floods of 1953, illustrating some aspects of dealing with widespread disaster, is given at Annex C.

Debriefing, inquiries and lessons emerging

2.32   In order to facilitate debriefing and to provide evidence for inquiries (whether judicial, public, technical or of some other format), it is essential that all events, decisions and actions taken are recorded. Each service, organisation or agency should maintain its own records. Chief officers and chief executives should consider instituting a formal debriefing system, on both a single- and multi-agency basis. Experience has shown that video recordings can be useful for these purposes.

2.33   Good record keeping serves a further purpose, whether or not there is a formal inquiry. It allows the lessons learnt from the disaster to be identified and made more widely available for the benefit of those who might be called upon to respond to other disasters in the future.

3 The immediate care and treatment of disaster victims

3.1 The care and treatment of those involved in a disaster lies at the heart of the response. This applies to the care and treatment not only of victims but of their relatives, friends and the rescue workers themselves who may be greatly affected by their experience.

3.2 Survivors or casualties may be thrown some distance away from the disaster site by the force of a crash or explosion. Other survivors may wander away in confusion. It is therefore important to consider the need to search the area surrounding a disaster. If this is necessary the task should be co-ordinated by the police incident officer. Where the task may be labour intensive and cover a wide area it may be appropriate to seek assistance from volunteers or the military, or both.

Care of uninjured survivors

3.3 Those who have survived a disaster uninjured (or with only minor injuries) may nevertheless be suffering from shock, intense anxiety and grief. They will, therefore, need to be treated with great sensitivity.

3.4 Survivors will often be able to provide crucial information about what happened and may be important witnesses at any trial or inquiry. A balance has to be struck between the requirement to gather evidence from survivors and the reluctance of some to remain at the scene of their distress. For example, prioritising information might help, so that only names and addresses are taken from those anxious to leave, with further details being obtained later. Information will usually be gathered at the survivor reception centre, a secure area to which all survivors who are uninjured or have only minor injuries should be taken.

3.5 Survivors' initial needs are likely to include support in their distress, food, drink, first aid to treat minor injuries and perhaps spare clothing and changing, washing and toilet facilities. They may also need immediate social and psychological support including help in finding temporary accommodation, in contacting family and friends, with transport back home and financial advice and assistance. It is possible that some apparently uninjured survivors will later display adverse symptoms. For this reason, medical and social services staff should be present at the survivor reception centre and, if necessary, at rest centres.

The injured

3.6 Injured survivors should be taken to a casualty clearing station (CCS) where medical and paramedical personnel will carry out triage and any appropriate stabilisation measures and ensure that casualties are evacuated in accordance with priority for hospital treatment. The CCS is usually sited in a building, tent or temporary shelter close to the ambulance loading point.

3.7 Responsibility for ensuring the proper establishment of medical communications on site, the transport of medical teams, the distribution and replenishment of medical and first aid supplies, the provision of all ambulance resources necessary for the ongoing treatment of casualties and (in liaison with the medical incident officer) conveyance of casualties to the receiving hospital(s) all lie with the senior ambulance incident officer. Where appropriate, this officer will also have responsibility for making arrangements to take casualties to distant specialist hospitals by helicopter.

Fatalities

3.8 The authority of HM Coroner is required before those who have been pronounced dead can be removed; arrangements are then the responsibility of the police. Until a doctor has pronounced life extinct and labelled the body accordingly, bodies (or parts of bodies) must not be moved except to preserve them from destruction by fire or chemicals or to rescue or provide medical treatment for trapped survivors. If it is necessary to move bodies, both they and the positions from which they were moved should be marked appropriately to provide those investigating the circumstances of the deaths with the necessary information about the bodies' positions at the time of death.

3.9 The usual practice is for bodies to be taken to a temporary mortuary where autopsies can be carried out, although they might be taken first to a temporary holding area. The ambulance service may assist in removing the dead from the scene. The need to preserve evidence at the scene of a disaster – including that identifying the deceased – must be borne in mind. For this reason, valuables, which can help in identification, should not be removed from bodies for safekeeping.

Casualty bureaux

3.10 In the event of a disaster the role of the police casualty bureau is to provide a central contact point for all those seeking or providing information about persons who might have been involved and to collect data and collate all records. As part of this process the police may send documentation teams to each receiving hospital, the mortuary and the survivor reception centre. The functions of the bureau are:

a. handling enquiries from the general public about relatives and friends who might have been involved;

b. collating details of survivors, their condition and their whereabouts;

c. informing enquirers of the condition and whereabouts of the survivors;

d. confirming areas of evacuation and the location of evacuees;

e. gathering data to assist in the identification of casualties; and

f. compiling a list of persons believed to have been involved who are now missing.

3.11 Once the casualty bureau has been established, its telephone number(s) will be publicised through the media, with the public being asked to provide information on persons thought to have been involved in the disaster who have not been accounted for. This information assists the police in their task of identifying casualties and the deceased.

3.12   The task of identifying the dead is carried out by the police identification commission, which is overseen by the police incident officer. The functions of the identification commission include liaison with, and compiling identification evidence for submission to, HM Coroner and controlling the running of the mortuary. When a positive identification has been made, it is the responsibility of the police to inform the next of kin and this would be done in person.

3.13   In the event of an air crash, the Emergency Procedures Information Centre (EPIC) managed by British Airways at Heathrow may open. If so, its telephone number will be broadcast. EPIC would act as a central airline information co-ordinating point, collating information about reservations, next of kin and other relevant data, and would work in support of the police and HM Coroner.

3.14   If foreign nationals have been – or are thought to have been – involved in the disaster, the Foreign and Commonwealth Office will refer any enquiries from foreign consulates, embassies or high commissions to the casualty bureau. The police will, in accordance with the Vienna Convention on Consular Relations, inform the consular authorities of the death of any of their nationals. The ambassador or high commissioner may wish to visit the disaster scene. Arrangements for VIP visits are discussed in the next chapter.

Friends and relatives

3.15   Recent experience of disasters has shown that, if they believe their friends and relatives may have been involved, it is likely that many people will travel to the scene. If necessary, a reception centre for friends and relatives will be established by the police (usually in consultation with the local authority and commercial, industrial or other organisations concerned) and staffed by the police, local authority and suitably prepared voluntary organisations, including representatives of churches and other faith groups. The fullest possible information should be given to enquirers seeking news of those involved in a disaster. Friends and relatives, who may be feeling intense anxiety, shock or grief, need to be treated with sympathy and understanding. Access to the reception centre should be controlled to prevent those inside being disturbed by media representatives or onlookers.

Evacuation and shelter

3.16   In some circumstances it may be necessary to advise the public on whether they should evacuate a given area or stay put and shelter indoors. Such circumstances include risks to life or health from:

a. the release or threatened release of radioactive materials;

b. the release or threatened release or other hazardous substances;

c. the spread of fire;

d. explosion;

e. severe storms;

f. flooding;

g. earthquake;

h. environmental contamination.

3.17 The possible need for evacuation in the event of the release or threatened release of radioactive material is addressed in 'Arrangements for Responding to Nuclear Emergencies' published by HMSO on behalf of the Health and Safety Executive (ISBN 0 11 885525 5) and so is not considered further in this chapter.

3.18 In the event of the release or threatened release of non-radioactive hazardous materials, additional information on the nature of the risk may be obtained from the fire service, from chemical data systems and other accredited sources. One crucial factor in determining the area to be evacuated will be the forecast speed and direction of the wind which, together with other 'Chemet' advice, can be obtained from the appropriate Regional Weather Centre.

3.19 Warnings of severe storms or other adverse weather are issued by the Meteorological Office and/or Regional Weather Centres in the form of severe weather warnings. In addition, warnings of abnormally high tides that could possibly lead to flooding are issued by the Meteorological Office to the police and to the National Rivers Authority. The National Rivers Authority in collaboration with the police decide whether local flood warnings should be issued.

3.20 It is normally the police who decide whether or not to evacuate and define the area to be evacuated. Their decision will take account of advice from the fire service on risks associated with fire, contamination and other hazards, from the ambulance and social services on problems associated with moving people who are frail or disabled or at risk for any other reason and from the local authority on possible places of shelter within the area. The police, however, can only recommend evacuation and have no power to require people to leave their homes. Past experience has shown that people with domestic pets may be unwilling to leave their homes unless arrangements include their pets.

3.21 In considering or executing evacuation, care must always be taken not to put people at risk by bringing them outdoors when they might be more effectively protected by sheltering indoors. This is particularly important in the case of the release of hazardous substances.

3.22 The physical difficulties of large scale evacuation should not be underestimated. If it is decided to evacuate a given area, evacuation assembly points should be set up near the area and, if time permits, signposted. Those in the affected area should be advised to go to their nearest evacuation assembly point. This can be done by house-to-house calls or by using loud-hailers, mobile public address systems, radio or TV announcements or any combination of these methods. People taking prescribed and other medications should be reminded to carry these with them. At the evacuation assembly point, local authority officers should co-ordinate the dispersal of evacuees to reception and rest centres.

3.23 The emergency services and the local authority should, as far as is practicable, take steps to ensure the security of property left empty after evacuation. When arrangements are being made for evacuation and for securing property, attention must be paid to the safety of emergency service and local authority workers who might be exposed to risk whilst working outdoors.

3.24 Local authority staff should maintain a comprehensive index of evacuees and their whereabouts. This information will be needed by the police for casualty bureau purposes and later if it is necessary to interview witnesses. In order to account for all people evacuated from an affected area, it is important that those leaving reception or rest centres and intending to stay with friends or relatives are encouraged to register their eventual destination at the centre.

Social and psychological support

3.25 There may be an immediate need to provide social and psychological support to those who are suffering from the effects of the disaster, including members of the emergency services and others called upon to respond. There will undoubtedly be a need for social and psychological support services to be set up in the immediate aftermath and possibly in the longer term. Much will depend on the nature and scale of the disaster. Local authorities might find it helpful to refer to the guidelines for an action plan for local authority social service departments which are set out in Part 2 of 'Disasters: Planning for a Caring Response' published by HMSO on behalf of the Disasters Working Party (ISBN 0 11 321370 0).

Religious and cultural needs

3.26 Any disaster occurring in the UK may involve foreign nationals and members of ethnic and religious minorities, whether they are resident here or visiting from abroad. Emergency services, local authorities and others responding to the disaster should therefore bear in mind that survivors may not be fluent in English and may have particular requirements in terms of diet and medical treatment. Similarly, there may be particular requirements concerning the handling of the dead and arrangements for burial. Language problems and special dietary needs may also have to be taken into account at reception centres for survivors, friends and relatives and for evacuees at reception and rest centres.

4 Relations with the media

4.1   Recent years have seen a rapid advance in telecommunication and information technology capabilities. Television channels devoted entirely to news and extended news programmes on other channels are with us to stay. The impact made at the scene of a disaster by those engaged in gathering material for the media can be massive and it is vital to prepare for the influx of media representatives – local, national and international. In little more than 12 hours after Pan American Airways flight 103 crashed there were at least 500 media people at Lockerbie, including 17 TV crews and a vast array of equipment – vehicles, cameras, communications and so on.

4.2   The purpose of this chapter is therefore to highlight some of the problems which can arise in handling the media's voracious appetite for information in the event of a disaster and to suggest arrangements for overcoming those problems. It is also worth noting too that certain incidents, such as the Hungerford shootings in 1987, may not necessarily be regarded in this context as disasters but will nevertheless generate a high level of ongoing media interest, posing similar difficulties.

Assisting the media

4.3   In the first instance, the load in coping with media pressures usually falls to the police in their role as co-ordinators of the response at and around the scene of disaster and with their responsibility for criminal investigation. But there are other aspects of a disaster – temporary accommodation for victims and perhaps their relatives and friends, safety of damaged buildings, road access and so on – which would call for a quick reaction by local authorities and so they too must be involved in the media response from the outset. Depending on the nature of a disaster, attention may also focus on industrial operators and commercial or other organisations; for example, Occidental Petroleum (Caledonia) Ltd was in the front line of media attention after the Piper Alpha disaster.

4.4   The guidance in this chapter is offered primarily to the emergency services, in particular to chief police officers in their role as co-ordinators at the scene, and to local authority chief executives. However, it is hoped that the guidance will also be of use to other organisations which may become involved in the disaster response.

Initial actions

4.5   Media personnel will arrive very quickly. They will often have learnt of the disaster at the same time as the emergency services and, when they arrive, will expect to have access to the facilities they require. They will also expect an instant response to their requests for information and briefing. Demands from local and regional media will quickly be augmented by demands from national and – depending on the nature of the disaster – international media.

If these demands are not anticipated, media representatives are likely to add to the confusion.

4.6 Experience has shown the value of immediately dealing with the following points:

a. control of access to the disaster site. This is a police responsibility, put in place whenever practical, intended to allow rescue services to carry out their work unhindered and to preserve evidence at what may be the scene of a crime. It must be anticipated that the broadcasting media in particular will bring large vehicles to the scene. In addition, helicopters are often deployed and control of the airspace may be an early consideration (see paragraph 2.16).

b. establishing a media liaison point. This is a designated point at the disaster scene, preferably outside the outer cordon, for the reception of media personnel, accreditation (see paragraph 4.11) and briefing on arrangements for reporting, filming and photography. It may be little more than a rendezvous point with further facilities provided at a media centre as described at paragraphs 4.15–4.17.

c. nominating a media spokesperson. The swift attendance at the scene of a spokesperson (likely to be from the police) should ease the pressure from the media. Failure to arrange this will prompt media representatives to approach anybody available, which could lend credibility to inaccurate sources. While the spokesperson takes the load off those coping with the disaster, brief interviews with the senior police, fire, ambulance or other officers at the site will add authority to the information being given.

d. assistance from the Central Office of Information (COI). Consideration should be given to seeking reinforcement from one or more experienced COI press officers. (See also paragraph 4.15).

4.7 The services involved are likely to receive various approaches from the media. Lack of co-ordination and information-sharing will result in different, and even contradictory, messages emerging. It is therefore important that the media response is co-ordinated through the nominated spokesperson from the earliest possible stage.

4.8 In addition, at this most difficult initial stage of the disaster response, it may be helpful for the following points to be borne in mind:

a. the media may need to be reminded that in the period immediately following a disaster, no-one can know precisely what has happened. Initial statements should focus on what is happening, what the limitations of knowledge are at the time and what is being done to arrive at a fuller appreciation of the situation. If such statements are backed by a commitment to provide accurate information as soon as it is available, media personnel are more likely to attend briefings and thus accept a measure of control, particularly if the briefings are scheduled at regular intervals.

b. meanwhile, any factual statements – particularly from emergency services' eye witnesses – will be welcomed. However, such statements should not include speculation on the cause of the disaster, nor premature or uncorroborated estimates of the numbers of casualties.

c. care should be taken that information about casualties is not released until details have been confirmed and the next of kin informed. It may be necessary to explain that it can take a long time to identify victims.

d. limitations on the release of information should be frankly explained, particularly the need not to prejudice what may become a criminal prosecution.

4.9 There will be great pressure on reporters to seek interviews with survivors and relatives but many will feel too shocked and distressed to give interviews. The first consideration should always be the well-being of the individual. It does, however, relieve pressure on all concerned if a willing and able survivor, relative or friend agrees to speak at a press conference to characterize the disaster for all. The Broadcasting Standards Council's monograph 'Survivors and the Media' by Ann Shearer (John Libbey and Company Ltd, 1991) is recommended to all those who might be involved in the media response to disaster. In addition, the Press Complaints Commission is administering a self-regulatory code of conduct drawn up by the newspaper industry. The code deals with many types of journalistic abuse which may cause particular stress at the time of disaster, including harassment, invasion of privacy, intrusion into grief and shock, unwanted involvement of relatives and friends and interviewing or photographing children.

4.10 It is important to develop the best possible relationship with media personnel from the start. Pressure of competition between media teams and individuals will make them sensitive to any restrictions which appear to them to be unfair. If they feel they have been treated unreasonably, some representatives of the media will try to make their own arrangements which may obstruct rescue work and compromise evidence for any necessary investigation. Well managed media relations should alleviate these problems and should also allow positive advantage to be taken of the help which the media can provide, for example by broadcasting messages asking people to avoid the disaster area, publicising details of any evacuation planned and broadcasting casualty bureau telephone numbers.

4.11 When circumstances permit, police incident officers may wish to consider whether there should be on-site facilities for issuing passes to media personnel or otherwise accrediting media representatives. Any system adopted should allow rapid identification of those allowed access to certain areas.

4.12 The management of large numbers of media representatives can be assisted through pooling arrangements, particularly if safety or security considerations restrict access to a disaster site. A pool might, for example, comprise one television crew, one newspaper reporter, a stills photographer and a radio reporter. Although a limit can be set on the number of media personnel to be allowed access to the site, such restrictions are seldom welcome and it is best to allow the media to decide who their representatives should be. Additional pooling facilities may be required for overseas media representatives. It is helpful to identify suitable sites for coverage of the disaster by media personnel who are not, at the time, included in pooling arrangements.

A media response organisation

4.13   In the event of a disaster, the initial actions just described may be no more than holding arrangements. As events develop, the need for a comprehensive media response organisation headed by someone who equates to a public relations manager may become apparent. It is recognised that not all organisations or agencies have such a person, but experience has shown it to be extremely helpful. The public relations manager for the disaster can be from the police or appropriate local authority, depending on circumstances and locally agreed plans.

4.14   It is essential for the public relations manager to be fully involved in the senior management arrangements for the disaster (see paragraphs 2.21–2.24), for example by attending senior level meetings, so that he or she is fully in the picture and can plan the media response. It is recommended that the public relations manager oversees all aspects of the media response, including:

a. activities at the media liaison point;

b. arrangements for the media to visit the site, possibly including transport where the disaster is in a remote area;

c. accreditation of media personnel;

d. management of the media centre when one is established (see paragraphs 4.15–4.17); and

e. monitoring of likely media activities related to the disaster but at locations remote from the disaster scene (see paragraph 4.21).

Media centre

4.15   A disaster may also justify the establishment of a media centre to provide working accommodation for media personnel, a news conference and briefing area, facilities for monitoring television, radio and newspapers and a press office with communications equipment. The media centre may be set up by the police or by the local authority; responsibility for establishing the centre should be agreed in advance as part of the planning process. It should wherever possible be close to the scene of the disaster and staffed by representatives of the major organisations with responsibility for handling the disaster; these should be people familiar with media relations – press officers wherever possible. Consideration should be given to inclusion of government press officers (see paragraphs 4.22–4.24) and seeking assistance from the regional Central Office of Information which has expertise in setting up media centres and also has resources to help run them.

4.16   Chief police officers and local authority chief executives may have identified buildings in their emergency plans which could be used as a media centre, although experience has shown that facilities can be set up rapidly in a suitable building not previously designated as such. An initial, temporary option is to use a large and suitably equipped vehicle, for example an adapted police command vehicle. This can provide limited facilities but will not accommodate the numbers of media personnel likely to be covering a disaster.

4.17   A media centre offers a number of advantages to all concerned:

a. it provides the media representatives with a known source for the most accurate and up-to-date information which the authorities can make available;

b. once links with the rescuers and others central to the disaster response are in place and spokespersons have been nominated, smooth flows of information can soon be established, compared and co-ordinated;

c. there is then a better chance of identifying and dealing with any potential differences in approach and agreed approaches can be quickly relayed to the emergency services and other control centres;

d. oversight from the media centre should help to control media presence at the site, whether this is for photo-opportunities or briefing;

e. monitoring arrangements may be set up at the media centre to enable all concerned to be aware of what the media are broadcasting or publishing;

f. in the event of a widespread or multi-site disaster, a single media centre could serve as a focus for several media liaison points at differing locations; and

g. the same media centre may suffice for both initial and recovery phases of the disaster response.

Transfer of the media operation to local or health authority control

4.18   It has already been noted that the police are likely to take the lead initially in dealing with the media in their role as co-ordinators of the disaster response. As the emphasis switches to the recovery, the overall lead may pass to the relevant local authority, who might then take the lead in coping with continued media interest and providing any necessary public advice and information. Close and continuing co-operation between the police and local authority media teams from the outset will achieve a smooth handover.

4.19   Experience has shown that local authority arrangements must be agreed between the local authorities involved to avoid the risk of contradictory messages emerging. The police should also be consulted.

4.20   In some circumstances—when, for example, after the initial phase of the disaster response the focus of media interest moves to NHS hospitals–it may be appropriate for the local health authority to take control of the media operation. Again, a handover will be facilitated if the police (particularly the casualty bureau) and NHS media teams have been working in close cooperation from the outset.

Remote handling

4.21   In some disasters, attention has focused on communities and individuals living many miles from the scene who nevertheless become a centre of media attention. An example is the home town of those killed whilst travelling. This, too, may require co-ordinated media-handling arrangements to ensure an efficient and coherent response from the authorities, which may need to include measures to protect people from excessive media attention.

Liaison with central government

4.22   A disaster brings with it calls for ministerial statements to Parliament and the media. It is the responsiblity of the lead department to co-ordinate a consistent and properly considered response by government (see Chapter 6).

On media matters, therefore, chief officers of police and local authority chief executives should look to the lead department's press officers as their main central government contact at the scene.

4.23 In practice, government press officers may need to be quickly at the scene in order to:

a. explain the involvement of government or the relevant regulatory body in the disaster response;

b. be involved with any arrangements for VIP visits (see paragraphs 4.25–4.30); and

c. provide feedback to central government.

4.24 When central government accident investigators are called in, the relevant government department's press office's involvement may continue far beyond the initial stages of the disaster.

Visits by VIPs

4.25 A government minister may make an early visit to the scene of a disaster and the injured, not only to mark public concern but also to be able to report to Parliament on the disaster response. If a government minister is to visit the scene, he may be accompanied by the local Member(s) of Parliament; this would be arranged by the minister's private office. It is possible that the scale of a disaster may in addition prompt visits by a member of the Royal Family and/or the Prime Minister. Local VIP visitors may include the Lord Lieutenant, the High Sheriff, religious leaders, local MPs, mayors, chairpersons and other elected representatives. If foreign nationals have been involved in the disaster, their country's ambassador, high commissioner or other dignitaries may also want to visit the scene.

4.26 Visits to the scene need to take account of the local situation and the immediate effects of the disaster on the local community. It may be inappropriate for VIP visitors to go to the disaster site itself whilst rescue operations are still going on, particularly if casualties are still trapped. VIP visits should not interrupt rescue and life-saving work and the police, as co-ordinators of the disaster response, should be consulted about the timing of visits.

4.27 VIP visits will inevitably cause some disruption, and visitors will want this to be kept to a minimum. However, there are also dividends to be gained from such visits as they may boost the morale of all those involved, including the injured and the emergency services, and give an opportunity to place on record public gratitude for what has been done.

4.28 Local services are, of course, experienced at handling VIP visits in normal circumstances and many of the usual considerations will apply to visits to a disaster site. It may be necessary to restrict media coverage of such visits, in which case pooling arrangements may be made (see paragraph 4.12).

4.29 Visiting ministers and other VIPs will probably require briefing before visiting the site (which they will doubtless want to see) and will certainly require briefing before any meetings with the media. In the case of government ministers, arrangements should be agreed with the government press officer.

4.30 In addition, VIPs are likely to want to meet those survivors who are well enough to see them. It will be for the hospital authorities to decide, on the basis of medical advice and respect for the wishes of individual patients, whether it is appropriate for VIPs and/or the media to visit the casualties. If the media cannot have access to wards, the VIP can still be interviewed afterwards at the hospital entrance about how patients and medical staff are coping.

Sustainability

4.31 Disasters place enormous demands on all involved in the response. Media interest, particularly if it is international, can create pressure throughout a 24 hour period. Chief officers of police and local authority chief executives will wish to take sustainability into account during the response to a disaster and seek mutual aid accordingly. The pooling of resources in a joint media centre should be helpful in this respect.

5   The Voluntary Sector

Introduction

5.1   Disasters can overstretch the resources of the emergency and local authority services and the value of additional support from the voluntary sector has been demonstrated on many occasions. These occasions have also shown that voluntary activity is more effective if planned for in advance. This chapter describes the principles to be followed.

5.2   The voluntary sector in the UK is large and often well organised. Those preparing disaster responses should take into account four kinds of voluntary effort:

a. established organisations such as the British Red Cross Society, St John Ambulance, Salvation Army and the Women's Royal Voluntary Service;

b. specialist skills offered by, for example: groups of doctors (such as the British Association of Immediate Care), voluntary radio operators (such as the Radio Amateurs Network, RAYNET) and search and rescue organisations including the Royal National Lifeboat Institution and cave and mountain rescue teams;

c. individuals, not necessarily in recognised voluntary organisations, whose help is offered or requested on the day; and

d. organisations which specialise in providing emotional support such as Cruse and the Samaritans.

5.3   Examples of the wide range of activities which can be undertaken in support of the statutory services are set out in Annex D to this guidance.

Planning the voluntary sector's contribution

5.4   General issues of planning and preparing the disaster response are discussed in Chapter 7. But in planning how the voluntary sector might contribute, the following points are particularly relevant:

a. the skills and expertise available from the voluntary sector will vary from place to place. In establishing what is available locally the statutory authorities should not overlook voluntary wartime skills in addition to those normally associated with the disaster response. A record of available local voluntary resources should be made and kept up to date;

b. the voluntary sector must be able to demonstrate its capabilities to the statutory services. It is of vital importance that, if disaster strikes, the voluntary sector should be able to contribute what has been mutually agreed and written into local plans. Voluntary agencies must therefore be able to demonstrate that their support is reliable, consistent and sustainable to the required standard;

c. the statutory and voluntary sectors should be clear about their respective roles in a disaster. Voluntary organisations must appreciate that the statutory services bear the responsibility for the overall disaster response but, equally, the statutory services must develop an understanding of the voluntary organisations' own structures and constraints; and

d. there should be agreement as to how, on the day, voluntary effort will be called out, what it will do, how it will be organised and how volunteers will be identified.

5.5 Planning will be more effective if it is carried out with the voluntary sector within a well-defined and mutually agreed structure kept up-to-date by regular formal and informal contact at a local level. One possible model, in which voluntary organisations are grouped on the basis of functions and linked to the statutory authority responsible for those functions, is described at Annex E to this guidance. But there are many other possibilities; the essentials are that there should be a structure which suits local circumstances, which is understood by all concerned and that points of contact are clearly identified.

5.6 When disaster strikes, people who are not part of any voluntary organisation will also wish to help and will offer their services. The local authority, in conjunction with the police, should establish a volunteer receiving point away from the scene to deal with such approaches. If the volunteer receiving point wishes to accept the offer of help, the volunteer can then be asked to report to the rendezvous point or to any other designated position. It should be noted that individual volunteers may require a considerable amount of management and leadership if they are to be effective: experienced members of established voluntary organisations may be able to help with this.

Training and exercising

5.7 General issues of training and exercising are also discussed in Chapter 7. When considering the particular contribution which the voluntary sector can make to the disaster response, it should be noted that established voluntary organisations and volunteer groups will usually have their own arrangements for training and exercising. Wherever possible these should be demonstrated to the local statutory services. Additionally, the statutory services and voluntary agencies should aim for joint training and exercising so that problems can be identified, plans and procedures updated and working relationships fostered.

Action following a disaster

5.8 Plans should include provision for voluntary organisations to be alerted or called out by, or at the same time as, the local statutory services or authorities they support; voluntary organisations should not normally attend the scene unbidden. Where volunteers are asked to go to the scene they should normally be asked to assemble at the rendezvous point described in paragraph 2.12d before being assigned to tasks. Where there are a number of voluntary organisations providing support within the same or related functional areas, a 'cascade' call-out system may prove useful.

5.9 A prolonged disaster may require voluntary organisations to operate a shift system. In planning for this eventuality the use of mutual aid from neighbours or from further afield should be considered.

6 Central government and lead departments

The lead department concept

6.1 It is fundamental to the arrangements for dealing with disasters in the UK that the first response is at the local level. Where local services find that the scale of a disaster puts it beyond the capacity of their own resources, their recourse is usually to mutual aid arrangements with services in adjacent areas.

6.2 More often than not, however, central government will also have a role to play. This may be an active role—where, for example, local services seek specialist advice or assistance from a central government department or the main source of information about the disaster lies at central government level (eg in the case of overseas nuclear accidents or satellite accidents). On other occasions the central government role may be limited to dealing with parliamentary, media and public enquiries. In either case, a specific government department will be nominated to take the lead.

6.3 The nomination of a lead department does not affect the underlying principle that wherever possible day-to-day procedures and links should be used for dealing with disasters. Local organisations should therefore aim to use normal links with government departments rather than channelling all dealings through the lead department. If, exceptionally, special arrangements need to replace these normal day-to-day links (for example to co-ordinate public information as described in paragraph 4.22) this would be made clear to all concerned.

Nomination of the lead department

6.4 Responsibility for ensuring that a lead department is nominated in good time to respond to an emergency rests with the Cabinet Office, as part of its normal role in co-ordinating activities which involve a number of government departments. The work is carried out by the secretariat of the Civil Contingencies Unit (CCU), a group of ministers and officials which meets when necessary under the chairmanship of the Home Secretary to consider the handling of emergencies. The CCU secretariat will remain in close contact with the nominated lead department throughout the emergency and arrange meetings of ministers and officials whenever that is required.

6.5 The arrangements for deciding which is the appropriate lead department are regularly reviewed by the CCU secretariat; the aim is to ensure that there are clear criteria for nomination so that departments know in advance the circumstances in which they will be expected to take the lead and can plan ahead and be ready to move into action immediately. The choice of lead departments is based on:

a. *The nature of the disaster.* There is usually a clear link between the nature of a disaster and the normal business of the department.

b. *Access to information.* Ready access to a flow of up-to-date and accurate information is essential for a lead department. The chances of achieving this are greater if officials of the department, agencies and organisations concerned know each other and have worked together. The public information division of the lead department has a special responsibility for ensuring that information originating from central government is consistent with that being given locally.

c. *The availability of facilities.* Where a disaster is large-scale or protracted, it may be necessary to use a dedicated emergency room and communications. Those departments most likely to be called upon to act as lead departments have such facilities.

Departments currently nominated to act as lead department for the different categories of emergencies specified are set out in Annex F. If the circumstances do not exactly fit these categories, the CCU secretariat decides which department will take the lead.

6.6 Circumstances may change as a disaster develops or as the recovery phase is entered and it may become appropriate for the department originally nominated to lead to hand over the task to another department. This happened in the aftermath of the storms in October 1987 when the Home Office took the lead initially but handed over to the Department of the Environment at a later stage. The CCU secretariat would ensure that any such change was agreed interdepartmentally before being implemented and notified to everyone concerned.

The role of the lead department

6.7 A central government lead department is expected to be prepared to undertake some or all of the following tasks:

a. Co-ordination of the activities of central government departments in the response to a disaster, providing a framework within which individual departments can discharge their specific responsibilities. (An important part of this work will be to ensure that the necessary links are established with the local response.)

b. Co-ordination of the collection of information on the disaster and its effects for the purpose of

i.   briefing ministers,

ii. informing Parliament, and

iii. providing information to the public and the media at national level.

6.8 The way in which a government department performs its lead department role will depend on the circumstances. In straightforward cases no special arrangements may be necessary: lead department officials will work from their normal offices or, out of normal working hours, from home.

6.9 Where circumstances demand, the lead department would activate special procedures, such as opening an emergency room, and may need to arrange regular meetings of the relevant departments (either directly or through the CCU secretariat).

6.10 However these activities are organised, it is incumbent upon the lead department to ensure that there is no unjustified duplication in requests for information from those busily engaged at the scene. Departments are expected first to check whether information they require for their own purposes is already available within the lead department. In a great many cases the lead department will be able to obtain information collected by other departments for their own use, avoiding the need to trouble people at the scene.

The territorial departments

6.11 If a disaster affected only Scotland, Wales or Northern Ireland the respective territorial department would, where appropriate, take the lead. Even if not nominated to lead, the territorial departments would have an important role in those parts of the UK for which they are responsible, acting in close co-operation with the nominated lead department. (One example is the crash of Pan American Airways flight 103 in December 1988. The Department of Transport was in the lead, but the Scottish Office played a major role in co-ordinating central government action in Scotland in support of the combined response based at Lockerbie).

Regional Emergency Committees (RECs)

6.12 Contingency arrangements are in place under which ministers may activate what are known as Regional Emergency Committees (RECs) should it be felt that the additional stresses caused by a disaster require support to be given to the normal administrative machinery at the local level. RECs are not an additional level of government; they are an integral part of central government standing ready to operate when required. They have no executive authority other than that delegated to departmental representatives by their parent departments.

6.13 In England, RECs (which are based for convenience on existing Home Defence regions) are chaired by the Regional Directors of the Department of the Environment. In Scotland, Wales and Northern Ireland a senior official of the respective territorial department chairs an equivalent committee. Other departments are represented by officials from the appropriate regional office or from headquarters. The emergency services are normally represented, as are other relevant organisations such as British Telecom. Representatives of other organisations may be co-opted to the REC as required.

6.14 The principal functions of the RECs are:

a. to provide an overview of the situation in their region and see that comprehensive and co-ordinated information passes between central government and local authorities and services; and

b. to assist in determining priorities for scarce resources and, as far as possible, resolve local problems.

Inspectorates

6.15 There exists a range of inspectorates, each with its own specific duties and responsibilities during normal times—for example, Her Majesty's Inspectorates of Constabulary and of Fire Services and the Inspectorates of the Health and Safety Executive. Many of these would have particular roles to play in the event of a disaster, often at the scene.

6.16 Liaison with the inspectorates at central government level will be channelled through the parent department. Officials of that department, and often a representative of the appropriate inspectorate, are likely to be present in the lead department's emergency room if one is set up.

7 The combined approach to planning, training and exercising

Introduction

7.1 The foundations of the combined response are the preparedness and expertise of each of the local emergency services and of the appropriate departments within the local authorities. The effectiveness of the response turns on these being brought together and linked with the roles which industrial or commercial concerns (including utilities), voluntary organisations and others may be called upon to play. At the core of local preparedness, however, is the framework which the emergency services and local authorities have evolved for planning, training for and exercising the local response. This chapter offers guidance on these subjects.

The combined approach to preparedness

7.2 Effective preparation of the disaster response requires co-operation between chief officers and their staffs and such co-operation is itself likely to be more effective if it is properly structured. Various structures are possible, but one which has been adopted in a number of areas is the establishment of a police-led senior co-ordinating group in which chief officers of the emergency services, the local representative of NHS management and local authority chief executives meet at intervals to consider such matters as plans, procedures, joint training and exercises. Much of the rest of this chapter is based on this arrangement but, whatever the arrangement, it is the principles which are important. Structures need to be evolved to meet local circumstances; for example in metropolitan areas (and in the case of some counties) arrangements will need to recognise that local authority boundaries do not match those of the emergency services.

7.3 The senior level co-ordination arrangements may need to be supported by working level groups to carry out detailed work. These lower level arrangements may involve personnel whose job is primarily concerned with disaster preparedness, for example emergency planning officers and their emergency services equivalents.

Planning

7.4 The senior co-ordinating group or its equivalent will want to ensure that local plans are properly co-ordinated and cover all reasonably foreseeable local hazards. It is good practice to involve as necessary in this work others whose responsibilities are particularly relevant to local disaster preparedness, including statutory bodies such as the National Rivers Authority and industrial and commercial organisations. The Health and Safety Executive may also have a role to play in providing hazard and risk information and acting as a source of local expert knowledge. Such an input may be of value in ensuring that preparedness covers industrial risks other than those for which there is a statutory planning requirement.[2]

---

[2]Official guidance on statutory planning for chemical and nuclear emergencies is issued by the HSE as follows:
- 'The Control of Industrial Major Accident Hazard Regulations 1984 – Further Guidance on Emergency Plans'. HS(G)25. Published by HMSO, ISBN 0 11 883831 8.
- 'Arrangements for Response to Nuclear Emergencies' (referred to in paragraph 3.17).

7.5   In addition to identifying and planning for specific hazards, the senior co-ordinating group or its equivalent may want to give thought to particular issues which have caused difficulties in the past. For example:

a. organisations involved in the disaster response will need to have co-ordinated arrangements for dealing with the media. Plans in this area might be prepared and updated in conjunction with the regional COI office and local media organisations. (See Chapter 4).

b. communications have sometimes proved to be inadequate in previous disaster responses. Attention should be given to the need for and means of inter-operability.

c. there is a need to ensure that different statutory services and authorities are not planning to use the same voluntary resources. To ensure that the voluntary sector's contribution is used to best effect, it may therefore be useful to jointly review plans for the use of the voluntary sector. (see Chapter 5).

7.6   disasters may well affect more than one local authority or police force area and the senior co-ordinating group or its equivalent might also include cross-boundary planning in its oversight of the combined response. Prior arrangements for mutual aid and some compatibility with the plans and procedures of neighbours are certain to be valuable in the event of a disaster which spans boundaries.

7.7   another area for consideration might be the need to sustain the response at an unusually intensive level, possibly over an extended period. Arrangements – probably involving mutual aid – should be devised to enable the response to be maintained, whilst at the same time recognising the need to relieve those engaged in it. In addition, account should be taken of the extent to which 'business as usual' must carry on in parallel with the disaster response.

Joint training and exercising

7.8   Planning should be underpinned by training and exercising and the senior co-ordinating group or its equivalent should oversee the multi-disciplinary aspects of these activities which, although inter-related, are considered in turn in the following paragraphs.

Subjects which may be included in joint training

7.9   For the police and fire services, and some of the NHS emergency services, responding to disaster can be considered as part of their day-to-day work. For others – local authorities and industrial or commercial concerns (including the utilities) – the response may be an extension of normal roles. For some groups, such as volunteers and military personnel, the response may be unrelated to normal day-to-day activities. There are, however, key training topics which are likely to be relevant to all who may be involved in the combined response. These are:

a. *awareness*. This is a universal requirement for all who might be involved, however remotely, in the disaster response. Attention should be drawn to the impact of disaster, to the numbers of victims which may result (including anxious or bereaved relatives and friends and indeed some of those respond-ing) and to the potential scale and complexity of the response. For the emergency services, awareness training will cover in more detail aspects of joint operations at and around the disaster scene. For local authorities, the emphasis will be on how departments can combine to support the emergency services in the initial stages of the response to care for those affected, to maintain essential services and to prepare to recover from the disaster. Such training might also include elected members who will need to be aware of the role they may have to play in a disaster. For other services a general familiarity with the combined response will suffice.

b. *liaison and co-ordination*. This should highlight the importance of liaison and co-ordination in the planning and delivery of the combined response. Good practice, conveyed through joint training, is crucial in improving the efficiency and effectiveness of the combined response.

c. *communications*. The importance of joint planning in communications has already been mentioned in the section on planning. Subjects for training might include:

– familiarisation with other organisations' forms, terminology and call-out arrangements where these are different; and

– information technology and information access.

d. *media relations*. In addition to the training required for preparation of the media response, practical training in interview and news conference techniques is likely to be of considerable benefit.

e. *recognition of stress*. Social and psychological aftercare following a disaster has already been mentioned in paragraph 3.25 and should be considered as part of the disaster response and recovery process. General training should be given in the recognition of stress both in the victims of disaster and in those responding and in appropriate referral of those suffering.

f. *the international dimension*. Included in this are:

– problems arising from disasters in which casualties include a large number of foreign nationals (see paragraph 3.14);

– European Community and other directives on issues such as the marking and transportation of hazardous material.

The need for training in this area is likely to increase as the effects of the introduction of the Single European Market are felt.

Types of training

7.10 Much training for the inter-disciplinary aspects of the combined response is likely to be carried out in-house but other sources are available. Some are listed below.

a. *Home Office Colleges*. The Emergency Planning College at Easingwold provides a range of interdisciplinary training in the form of seminars, workshops and courses. The Police Staff College, Bramshill provides a disaster management course primarily for senior police officers. The Fire Service College at Moreton-in-Marsh (which became an agency on 1 April 1992) currently provides some joint training courses.

b. *COI Courses*. As part of its regular training programme for Government Information Service officers, the COI offers one-day courses in specialist media liaison techniques for public relations officers and other officials who may be involved with the media at the scene of a disaster. The courses are designed primarily for government officers but are also available and suitable for officials from local authorities, emergency services and other organisations likely to benefit from such training. Participants are charged on a cost recovery basis.

c. *NHS/Medical*. At least one county-based ambulance service and the British Association for Immediate Care run courses on the command and control and the medical aspects of the disaster response.

d. *Academic institutions and other organisations* tend to run specialised courses. Information on some of these may be obtained from the Emergency Planning College, Easingwold.

Joint exercising

7.11   Joint exercising is a vital part of local disaster preparedness. Joint exercises test plans and procedures, provide practical training in the combined response to disaster and enable working relationships to be formed between those likely to be involved in that response. It is good practice for a combined response exercise schedule to be agreed locally.

7.12   Exercises may take a number of forms, including:

a. *Table-top exercises*. These can be particularly cost-effective. They can be designed to exercise the command and control of the overall combined response or any part of it. It is important in table-top exercises to be realistic and not to ignore the physical difficulties which, for example, deployments and communications present. These exercises are particularly useful because of the ability to 'stop-the-clock' and discuss points and decisions made as they arise.

b. *Computer-based simulation*. As various forms of information technology become established in operations rooms and emergency centres, this type of exercise may be particularly helpful in enabling managers and operators to practice in their operational environment. In addition, disaster response computer simulation is a technique which shows considerable promise.

c. *Live exercises*. These can be resource-intensive and should therefore be planned and prepared carefully. It is difficult to replicate the conditions of a real disaster fully; however live exercises are the best way (short of responding to a real disaster) in which to test procedures and liaison processes and to incorporate lessons learnt into revised plans.

7.13   There are a number of mandatory exercises run throughout the year by various agencies and full or partial participation in one or more of these may be helpful in rehearsing some or all of the local response. Examples include emergency exercises undertaken by:

–those operating nuclear installations and CIMAH sites

–HM Coastguard

– port and harbour authorities

– airport authorities

– British Rail.

It may also be possible to take similar advantage of non-mandatory exercises.

Definitions of major incident

Police/Fire Services

1. The following definition is set out in the Association of Chief Police Officers Emergency Procedures Manual and in the Fire Service Major Incident Emergency Procedures Manual.

'A major incident is any emergency that requires the implementation of special arrangements by one or more of the emergency services for:

a. The initial treatment, rescue and transport of a large number of casualties;

b. The involvement either directly or indirectly of large numbers of people;

c. The handling of a large number of enquiries likely to be generated both from the public and the news media usually to the police;

d. The need for the large scale combined resources of the 3 emergency services;

e. The mobilisation and organisation of the emergency services and supporting organisations, eg local authority, to cater for the threat of death, serious injury or homelessness to a large number of people.'

National Health Service (including Ambulance Service)

2. The following is extracted from Health Circular (90)25 'Emergency Planning in the NHS: Health Services Arrangements for Dealing with Major Incidents' – Department of Health 1990:

'There is no standard definition of a "major incident" which would satisfy the health service and other agencies likely to be involved, as each tends to determine such incidents in the light of its own responsibilities. For the purpose of this guidance however, a major incident arises when any occurrence presents a serious threat to the health of the community, disruption to the service, or causes or is likely to cause such numbers of casualties as to require special arrangements by the health service.'

Guidance addressed to emergency services, local authorities and other organisations

Service/ Department	Title	Issued by	Date
Police	Emergency Procedures Manual	Association of Chief Police Officers	1991
Fire	Major Incident Emergency Procedures Manual	Chief and Assistant Chief Fire Officers Association	1991
NHS	Emergency Planning in the NHS: Health Services Arrangements for Dealing with Accidents Involving Radioactivity	Department of Health	1989
	Emergency Planning in the NHS: Health Services Arrangements for Dealing with Major Incidents	Department of Health	1990
Ambulance	Ambulance Service Operational Arrangements: Civil Emergencies	Regional Ambulance Officers Group	1990
HM Coastguard	HM Coastguard: Operations, Planning and Procedures (CG3)	Department of Transport	1988
Local Authorities	Emergency Planning Guidance to Local Authorities	Home Office	Under revision
Local Authorities Emergency Planning Officers	Local Authorities Civil Emergencies Planning Manual	County Emergency Planning Officers' Society	1990
National Rivers Authority	Emergency Procedures for Flooding Incidents	Ministry of Agriculture, Fisheries and Food	1991

Service/ Department	Title	Issued by	Date
Pathologists	Deaths in Major Disasters: The Pathologist's Role	Royal College of Pathology	1990
Industrial Emergency Services Officers	Guide to Emergency Planning	Society of Industrial Emergency Services Officers	1989

Annex C

The East Coast Floods – January 1953

1. During Saturday 31 January 1953 freak winds drove a storm tide surge down the North Sea. By the evening this had reached the east coast of England where sea defences were over-topped and damaged by huge waves. Breaches occurred in 1200 places. With sea levels rising some 9 to 10 feet above normal high water marks, disastrous flooding occurred. Three hundred and seven people died and an estimated 40,000 were rescued, evacuated or fled from their homes. Twenty-four thousand houses, 200 industrial establishments and large tracts of land were flooded and there were severe losses of livestock and poultry. Worst affected were the five counties with their coastlines extending south from Lincolnshire to Kent.

2. The timing of the disaster – after dark on a cold, stormy night in the middle of a winter weekend – could hardly have been worse. Most urgent among the tasks tackled were the rescue of those stranded on roof tops and elsewhere, shelter and sustenance for those made homeless, repair of the sea defences to prevent further flooding at the next and subsequent high tides and the maintenance of supplies and services to communities badly affected but able to stay put.

Rescue

3. In the early evening the storm tide surge struck Lincolnshire and Norfolk. It moved south to strike Essex and Kent around midnight, drowning many asleep – particularly those in lightweight holiday chalets in coastal resorts. There are many accounts of self-rescue or rescue by neighbours. In addition, at the first signs of flooding, all available police were called out to assist. Joining rescue operations were the fire brigades, military personnel and the RNLI. Boats, amphibious vehicles, high chassis lorries and fire service equipment were variously pressed into action. Incident posts were set up at various points and were manned by the police and by liaison officers of other services. The incident posts were in communication with information rooms in police headquarters where rescue parties were co-ordinated. Requests for medical aid, ambulances and rescue craft and reports on the movement of displaced people were sifted and actioned. Police in the affected counties were able to sustain operations by means of reinforcements from the British Transport Police and from forces as far afield as Northumberland.

Shelter and sustenance

4. Particularly near areas where large-scale evacuations took place – for example Thurrock and Canvey Island – local authorities took the lead in setting up rest centres and in arranging temporary accommodation. They were strongly supported by voluntary organisations: the British Red Cross Society equipped, and helped in, many rest centres and the Women's Voluntary Service (WVS) staffed reception and rest centres and provided

37

large numbers of meals in them and elsewhere. The W V S distributed clothing to evacuees, many of whom were drenched in the freezing weather and unable to bring their belongings with them. The police were also involved in finding temporary places of refuge which were in some cases equipped by local welfare officers. Schools, chapels, drill halls, army camps, holiday centres and hotels were all used as reception and rest centres. Temporary accommodation was also provided through billeting arranged by the W V S and through offers of hospitality from friends and relatives as well as strangers. In association with rest centres, medical aid posts and emergency surgeries were arranged; food hygiene and sanitary services were maintained and temporary mortuaries were set up as necessary. With the help of voluntary organisations and local authorities outside the flood areas, the affected local authorities were able, with volunteer support, to carry out these emergency tasks promptly, thereby alleviating distress and preventing disease.

Repair of sea defences and river banks

5.   As mentioned earlier, the repair of sea defences was given high priority with a view to preventing yet further floods. The five river boards responsible had full-time civil engineering staff with specialised knowledge of sea defence work. Supported by their associated works organisations and with help and equipment provided by unaffected river boards outside the area, by fire brigades, by the military, and by many other organisations including firms in the Federation of Civil Engineering Contractors, the river boards set about their urgent tasks. These included first aid repairs to the damaged sea defences and clearing sea water as quickly as possible from certain flooded areas such as power stations and factories. At one time there were as many as 25,000 people working on these tasks in freezing winds under conditions of great difficulty and hardship. Hundreds of pumps, excavators, bulldozers and lorries were used and millions of sandbags were filled and placed in the worst breaches against future high tides and flooding.

Maintenance of services and supplies

6.   Very soon local authorities and water undertakings had to address the restoration of the water supply and sewerage systems. In many areas the water supply was contaminated by salt water, which had itself been polluted by contact with refuse tips and sewage works. At Sheerness the whole water supply was contaminated but it was decided to bring drinking water in barges from Chatham rather than evacuate yet more inhabitants. In other areas the water supply was maintained by emergency connection to other sources. At sewage works the flood water stopped electric pumping motors and interfered with the filters and sludge beds. The speed with which both water supply and sewerage systems were restored reflected great credit on the local and water authorities, who once again received invaluable help from authorities outside the flooded area, from fire brigades and Royal Navy divers.

7.   The 25,000 inhabitants of the Isle of Sheppey were completely isolated for a number of days. The Royal Navy kept them supplied and arranged a passenger service to and from the mainland.

National arrangements

8.   Reference has already been made to reinforcements from outside the flood area. In many cases these were co-ordinated from national headquarters.

9.   General Flood Headquarters was in the Ministry of Agriculture and Fisheries Land Drainage Division HQ in London. Linked to the river boards by teleprinter, the Ministry co-ordinated the repair of sea defences and directed supplies of materials to where they were most urgently needed.

10.   To co-ordinate the work of local fire brigades and organise reinforcements and reliefs, a temporary central fire control was set up in the Home Office with regional controls at Cambridge and Nottingham. Additionally, the Home Office Emergency Committee, comprising members of government departments concerned with disaster relief and representatives of certain voluntary organisations, met daily.

11.   National or regional HQs of the British Red Cross Society, St John Ambulance Brigade, the Women's Voluntary Service, the Salvation Army, the Church Army, the Royal Society for the Prevention of Cruelty to Animals – to name just some of those contributing – backed and co-ordinated the work of their local branches.

12.   Military assistance was provided on a large scale. In addition to the Royal Navy contributions already mentioned, some 11,000 army personnel were directly employed at the peak period on a range of tasks. These included the use of 1500 trucks, 50 amphibious vehicles, 65 heavy earth-moving machines, 300 boats of various sorts and material for 12,000 yards of improvised roads. Principal amongst the Royal Air Force contributions was the provision of air photography enabling a complete assessment of the flood damage and the identification of at least one breach in the coast defences unknown to authorities on the ground. The United States Air Force brought portable radios, amphibious vehicles and search lights to flooded areas and contributed notably to rescue operations.

Conclusion

13.   This brief account stops short of the massive recovery task which faced local authorities and others who rehabilitated communities and the environment and disbursed the Lord Mayor of London's national flood and tempest distress fund. Attention is drawn, however, to certain features of the initial response.

14.   *Rescue.* The police provided co-ordination through incident posts and force information rooms for many of the hundreds of rescue operations which took place. Fire brigades, military personnel and the RNLI were amongst those who made key contributions, providing liaison officers to the police incident posts as necessary.

15.   *Shelter and Sustenance.* Local authorities and voluntary organisations in the affected counties arranged shelter, sustenance, clothing and temporary accommodation for thousands of evacuees. Local authorities provided co-ordination for the alleviation of distress and prevention of disease in the difficult circumstances of flood damage and large scale evacuation.

16.   *Reinforcement.* The police, fire brigades, local authorities, voluntary organisations, river boards and water undertakings providing the disaster response were reinforced by colleagues from outside the affected area. In some cases this was arranged centrally in London. Through reinforcement it was possible to complete certain urgent tasks very quickly and also to sustain the response under adverse conditions.

17.   *Military Assistance*. The armed forces made a significant contribution to the response. They provided expertise, equipment and labour to help with rescue, evacuation, temporary accommodation, repair of sea defences, supply to isolated communities and the rescue and feeding of livestock.

18.   *National Support*. The alacrity with which reinforcement and military assistance were provided was representative of the spontaneous upsurge of sympathy and generosity from the British public. This was one consoling feature in a natural disaster without parallel in its effect on the country since Tudor times.

Examples of activities which can be undertaken by the voluntary sector in support of the statutory services

1. *WELFARE*

   Activities: Staffing reception and rest centres
   Feeding
   Provision of clothing
   Advice on entitlements, grants, loans, claims
   Resettlement of victims, evacuees etc
   Support and comforting

   In support of the Local Authority Social Services Department
   following statu- Local Authority Education Department
   tory services: Local Authority Housing Department and also
   (hereafter abbrevi-   police
   ated to 'In
   support of')

2. *SOCIAL AND PSYCHOLOGICAL AFTERCARE*

   Activities: Befriending
   Counselling
   Providing longer-term support

   In support of: Local Authority Social Services Department
   Local Authority Educational Psychologists
   National Health Service

3. *MEDICAL SUPPORT*

   Activities: Back-up Ambulance Service
   Medical aid posts
   Medical aid support in rest centres, emergency
      feeding centres etc
   Auxiliary roles in hospitals
   Welfare

   In support of: NHS Ambulance Service
   NHS Hospitals
   NHS Health Emergency Planning Officers

4. *SEARCH AND RESCUE*

   Activities: Mountain, cave, cliff, moor, inland waterways,
      coastal rescue

Supervision of other searchers (eg youth organisations)
Loan of equipment

In support of: Emergency Services[1]

5. TRANSPORT

Activities: Transport and escort of homeless, outpatients, next-of-kin, etc to and from airports, railway stations, hospitals, mortuaries, rest centres, hostels, etc

In support of: Local Authority Social Services Department
Emergency Services
NHS

6. COMMUNICATIONS

Activities: Providing radio communications
Vehicles
Messengers
Interpreters and translation

In support of: Emergency Services
Local Authorities
Utilities

7. DOCUMENTATION

Activities: Tracing – nationally and internationally
Assistance at Casualty Bureau in some local areas
Logging/diary procedures
Computer support (where available)

In support of: Emergency Services (especially police)
Local Authority Social Services
NHS

8. TRAINING AND EXERCISING

Activities: Analysis of training needs and capabilities
Devising instructional programmes
Joint planning and conduct of multi-agency exercises including call out arrangements and debrief
Formulation and dissemination of good practice

In support of: Emergency Services
Local Authority departments
National utilities
NHS

---

[1]Note – the emergency services may call on military assistance – particularly military search and rescue resources. Voluntary organisations may therefore sometimes find themselves working with military units.

Co-ordination of statutory and voluntary services

1.   Co-operation between the statutory services and local voluntary organisations can be facilitated by setting up a Voluntary Emergency Steering
Committee which could be co-ordinated by the county or local emergency
planning officer.

2.   One method of involving voluntary organisations in the planning for
disasters is to group them where appropriate on the basis of their functions
and linked with the statutory authority responsible for those functions. This
functional grouping can clarify the contributions which individual voluntary
organisations can make and enable statutory authorities and voluntary
organisations to maximise – and rationalise – the voluntary contribution. In
some cases there will be one statutory authority and one voluntary organisation linked to a specific function, for example H M Coastguard and the
Royal National Lifeboat Institution. In other cases, where a voluntary
organisation performs a range of functions, it would need to be associated
with more than one statutory authority and represented on all the relevant
functional groups. Annex D has shown functions which have been performed
in the wake of previous disasters, together with examples of the activities
involved and the statutory authorities normally responsible.

Lead government departments

Against the types of disaster or emergency listed in the left hand column below are shown the lead departments as nominated in June 1992.

1. FLOODING (coastal or riverine)

    Ministry of Agriculture Fisheries & Food or equivalent department of *Scottish and Welsh Offices* or the appropriate Government Department in Northern Ireland.

2. MARINE AND COASTAL POLLUTION (oil, chemical or gas)

    Department of Transport (Marine Pollution Control Unit).

3. RADIATION HAZARDS (arising within the United Kingdom)

    Department of Trade and Industry for civil nuclear installations in England and Wales; *Scottish Office* for civil nuclear installations in Scotland; *Ministry of Defence* for defence nuclear installations and nuclear material in transit; or *Department of Transport* for civil nuclear material in transit.

4. RADIATION HAZARDS (arising outside the United Kingdom)

    Department of the Environment.

5. NUCLEAR POWERED SATELLITE ACCIDENTS

    Home Office.

6. BLOW OUT ON OFF-SHORE INSTALLATIONS

    Department of Employment (Health and Safety Executive).

7. SEARCH & RESCUE (in UK Search and Rescue Region)

    Department of Transport (HM Coastguard) for civil shipping; *Ministry of Defence* for military shipping and aircraft, for civil aircraft at sea and for civil aircraft on land when the location is not known (when the location is known the emergency is treated as a transport accident – see 11 below).

8. SEVERE STORMS

    Home Office (or the appropriate territorial department) initially, with *Depart-*

ment of the Environment assuming the lead role at a later stage if appropriate.

9. OTHER GENERAL WEATHER EMERGENCIES

Department of Transport (or the appropriate territorial department).

10. DISASTERS OVERSEAS (in which UK assistance is sought)

Foreign & Commonwealth Office (Overseas Development Administration).

11. TRANSPORT ACCIDENTS (including those overseas involving UK registered ships and aircraft)

Department of Transport.

12. ACCIDENTS IN SPORTS GROUNDS

Home Office (or appropriate territorial department) in close consultation with the Department of National Heritage (Minister of Sport).

13. MAJOR EXPLOSIONS ARISING FROM LANDFILL GAS

14. DAM FAILURES

15. MAJOR STRUCTURAL FAILURES IN BUILDINGS (other than those caused by external impact, gas explosion, fire or industrial process)

(Note. If a government building is involved the relevant department would lead)

16. EARTHQUAKES

Department of the Environment or appropriate department of the *Scottish or Welsh Offices* or the appropriate Government Department in Northern Ireland.

17. SERIOUS INDUSTRIAL ACCIDENTS

Depending on the nature of the accident, lead responsibility would by taken by:

a. *Home Office* if the activities of the emergency services are the main focus of attention.

b. *Department of the Environment* if rivers, inland waterways (outside Port Authority jurisdiction) or water services in England are the main cause of concern, or gas clouds of toxic or unknown composition are threatening the environment or public safety.

c. *Ministry of Agriculture, Fisheries and Food* if the main focus of attention is safety of food supplies.

d. *Department of Employment* if the main focus of attention is the Secretary of State's responsibilities for the operations of the Health and Safety Executive.

e. *Scottish or Welsh Offices* or the appropriate Government Department in Northern Ireland for incidents in those parts of the UK for which their Secretaries of State are responsible.

Other possible, but less likely, candidates for the lead role could be:

f. *Department of Trade and Industry* if the main focus is on information about the industry involved and/or the economic implications of the disaster.

g. *Department of Health* if the main focus is on advice to the public on hazards and their effects or on activities of the NHS (ambulance and hospital services).

Glossary[1]

Ambulance Control Point	An emergency control vehicle providing an on site communication facility which may be at a distance from the incident scene and provides a focal point for NHS/medical resources attending the incident. Ideally the point should be in close proximity to the police and fire service control point vehicles (subject to radio interference constraints).
Ambulance Incident Officer	The officer of the ambulance service with overall responsibility for the work of that service at the scene of a major incident.
Ambulance Loading Point	An area, preferably hard standing, in close proximity to the Casualty Clearing Station, where ambulances can be manoeuvred and patients loaded.
Ambulance Parking Point	Place designated at the scene of a major incident where arriving ambulances can park, thus avoiding congestion at the entrance to the site or at the Ambulance Loading Point. These areas are also suitable for staff briefings, procurement of refreshments and restocking of equipment.
Cascade System	System whereby one organisation calls out others who in turn initiate further call outs as necessary.
Casualty Bureau	Central contact and information point for all records and data relating to casualties.
Casualty Clearing Station	An area set up at a major incident by the ambulance service in liaison with the medical incident officer, to assess, treat and triage casualties and direct their evacuation.
Chemet	A scheme administered by the Meteorological Office, providing information on weather conditions as they affect an incident involving hazardous chemicals.

---

[1]Definitions, where appropriate, are extracted from the relevant emergency procedures/planning manuals (see Annex B). The harmonisation of these definitions is being undertaken by a Home Office-led interdisciplinary working group.

Control Room	Centre for the control of the movements and activities of each emergency service's officers and equipment. Liaises with the other services' control rooms.
Cordon—Inner	Surrounds the immediate scene and provides security for it.
Cordon—Outer	Seals off the controlled area to which unauthorised persons are not allowed access.
Emergency Centre	Local authority operations centre from which the management and co-ordination of local authority incident support is carried out.
Evacuation Assembly Point	Building or area to which evacuees are directed for transportation to a rest centre.
Forward Control Point	The control point nearest the scene of the incident responsible for immediate deployment and security.
Friends and Relatives Reception Centre	Secure area set aside for use and interview of friends and relatives arriving at the scene.
Incident Control Post (Police and fire services)	The point from which the management of the incident is controlled and co-ordinated. The central point of contact for all specialist and emergency services engaged on the site.
Major Incident Procedures	Pre-planned and exercised procedures which are activated once a major incident has been declared.
Maritime Rescue Co-ordination Centre	H M Coastguard regional centre responsible for promoting the efficient organisation of search and rescue services and for co-ordinating the conduct of search and rescue operations.
Marshalling Area	Area to which resources and manpower not immediately required at the scene or being held for further use can be directed to standby.
Medical Incident Officer	Medical officer with overall responsibility (in close liaison with the Ambulance Incident Officer) for the management of medical resources at the scene of a major incident. He/she should not be a member of a mobile medical team.
Mutual Aid Arrangements	Cross-boundary arrangements under which emergency services, local authorities and other organisations request extra staff and/or equipment for use in a disaster.
Receiving Hospital(s)	The hospital(s) selected by the ambulance service (from those listed by the Regional Health

	Authority) to receive casualties in the event of any particular major incident.
Rendezvous Point	Point to which all resources arriving at the outer cordon are directed for logging, briefing, equipment issue and deployment.
Rest Centre	Building designated by local authority for the temporary accommodation of evacuees.
Statutory Services	Those services whose responsibilities are laid down in law: for example, police, fire and ambulance services, HM Coastguard and local authorities.
Survivor Reception Centre	Secure area to which uninjured survivors can be taken for shelter, first aid, interview and documentation.
Territorial Departments	Scottish Office, Northern Ireland Office and Welsh Office.
Triage	Process of prioritising the evacuation of the injured by the medical or ambulance staff at the Casualty Clearing Station.
Utilities	Companies providing essential services eg gas, water, electricity, telephones.

Bibliography

Military Aid to the Civil Community: A Pamphlet for the Guidance of Civil Authorities and Organisations Ministry of Defence 1989, third edition: AC 60421.

Instructions for Establishing Emergency Flying Restrictions Within the UK National Air Traffic Services 1989.

Arrangements for Responding to Nuclear Emergencies Health and Safety Executive: HMSO (ISBN 0 11 885525 5).

Disasters: Planning for a Caring Response Disasters Working Party: HMSO (ISBN 0 11 321370 0).

Survivors and the Media Ann Shearer (Broadcasting Standards Council Monograph): John Libbey and Company Ltd.

Control of Industrial Major Accident Hazard Regulations 1984 Further Guidance on Emergency Plans HS(G)25 Health and Safety Executive: HMSO (ISBN 0 11 883831 8).

Printed in the United Kingdom for HMSO
Dd296237 5/93 C20 G531 10170